Growing a Garden From Seeds

In Southwest Idaho

Learn to plant the right seeds at the right time

Laura Blodgett

Daily Improvisations Press
Marsing, Idaho

Table of Contents

Part One

Who Can Grow Plants From Seed?

I knew as soon as I did it that something was wrong. Though I had grown up helping my parents with the family garden in Oregon, this had never involved tulips. Now, with my first yard to garden in, albeit a rental, I had spontaneously purchased a potted tulip. The red tulip flower was in its prime. I had dug the hole, transferred the plant into the hole, and sat back on my heels to think whether or not I was forgetting something. I vaguely recalled something about cutting off flowers to stimulate more blossoms. So, I found my kitchen scissors and dutifully cut off that beautiful red flower.

As I stared at the perfect bloom now laying in the dirt, disconnected from its source, I felt a heaviness of disappointment settle in my chest. For some reason now I knew it would not grow another flower. I wasn't sure how I knew this now, but not thirty seconds ago. Thirty seconds ago would have been better. And why now did I have a remote recollection that tulips don't really grow in southern California, which was where I was at that time? I scanned my memory and the yards in the neighborhood. No tulips anywhere. Thus the final death knell rang for my tulip. It wouldn't even be growing again next spring.

Did this mean I had a "black thumb" or that gardening was some mystic art? I know that this isn't a story about growing from seed, but it is about learning to grow things. Anytime something is new to you, it is harder to problem solve. You don't have enough experience to easily figure out what is a big problem and what is just a minor set back. It is much easier to get discouraged.

Fortunately, my childhood experiences had at least left me with the idea that growing common garden vegetables from seed was something for everyday people. I could recall putting all kinds of seeds rather haphazardly in rows, as patiently directed by my parents. If seeds had grown into plants then, in spite of my childish techniques, surely this tulip incident was just a bump in the road.

About a three year bump, for that was when we moved to Idaho and bought our first house. There was a sunny location for my first official garden. I was also excited to see lots of tulips in people's yards! I was ready to try again and learn along the way.

I subscribed to a gardening magazine that inspired me with way too many ideas to implement, but I will always remember one piece of advice that both reduced stress and motivated me:

Just start your garden and enjoy it.
You can keep learning and making adjustments along the way.

In other words, my garden didn't have to be anyone's idea of perfect to be a legitimate garden.

To look at it from another angle, anyone who is willing to do the basics of tending, like planting and watering, can grow many vegetables from seed. You will learn how to fine tune that along the way and you will probably also develop skills that make it easier to grow more temperamental seeds, too. But you can grow a garden from seed.

Part Two

Why Grow Your Garden From Seeds

When I first started growing a garden, I thought everyone grew their vegetables from seed. I didn't know some people bought plants in pots and transplanted them into their gardens! When I did discover such plants, it felt like it would be cheating. It definitely seemed more expensive.

Since then I have re-evaluated and now I occasionally buy a plant or two for my vegetable garden. However, I still think growing my own plants from seed has many advantages:

1. I don't have to wonder exactly what varieties will be for sale at local garden nurseries. Thus, I can more easily and consistently be growing the plants I want instead of being at the mercy of what is stocked.

2. In my experience, there are more seed varieties available than plant varieties for most vegetables. This is simple economics. It takes more space and care to sell plants, so it is more work and riskier for the stores.

3. It is easier to make sure I get varieties that I might want for saving seeds. This is usually labeled on seed packets. Plants sold tend to be hybrids. *(The seeds produced by hybrid plants are not reliably like the parent plant.)*

4. I can be sure that my plants are well adjusted to my climate and environment from their start. *(Even if I use the greenhouse, which I will talk more about later)*

5. I theoretically get more plants for the same amount of money *(depending on germination rate, how I care for them, and money spent on seed starting supplies — more to come on that)*

There are obviously some advantages to just buying plants:

1. Less time and work is spent tending seedlings

2. No need for as much indoor space for germination.

3. You can potentially start off with larger plants.
(but that doesn't always give you a long term seasonal advantage, And you may still have issues with hardening them off.)

4. That's all I can think of, haha

Other things that discourage me from buying plants are:

1. It is not unusual to find they are root bound. *(they've been growing in the pots too long and their roots are compacted and tangled — they may do okay if the roots are stimulated with some rubbing and cutting to get them to grow out of the pot shape, but sometimes they don't recover quickly or at all from being root bound)*

2. The growth of their foliage has often been restricted. Plants that touch each other tend to grow taller and skinnier. This makes them stressed and more likely to get damaged.

3. Potted plants can introduce new insects and weeds into your garden.

4. And, to be completely honest, I feel somewhat disconnected from my garden if I'm not growing plants from seed.

So, if you are like me and want seeds to be part of your gardening, the next questions are:

1. Do I start seeds indoors or direct plant them outdoors?

2. When do I start seeds either way?

Both questions are related and I will be offering you many helpful guidelines. In the next section, let me give you some general things to consider, then after that I will go over particulars for growing many of the vegetables grown in SW Idaho. I have grown a lot of things from seeds, but even if I haven't grown something specific, I have probably germinated seed for something in that category, so pretty much every kind of vegetable should get covered one way or another.

Mostly I will talk about vegetable seeds, but I will also mention flowers that I have grown from seed, some of which are edible, such as nasturtiums and sunflowers.

When to buy your seeds

First, you don't necessarily need to buy all new seeds every year. How well any unused seed saves will depend on things like how you store it *(best in a cool and dark location)* and what kind of seed it is. Plus, you can always do a germination test of a few of the seeds in a damp paper towel inside a plastic bag to see how viable they are.

Seed catalogs start coming in the mail as early as December. I see seed in many stores as early as February. Obviously, the earlier you purchase it, the more there will be to choose from. Favorites do sell out.

Another reason I like ordering through a catalog is that it is more likely the seeds will have been kept in cooler temperatures. However, I am very diligent to check for packages in the mailbox and on the porch to avoid damage from moisture.

I will list some seed companies I have been happy with in an appendix.

Part Three

Deciding Whether to Start Seeds Indoors Or Direct Seed Outdoors

Just in case you are new to this, let me specify what growing indoor from seed means.

Starting seeds indoors typically means:

Planting seeds in pots under conditions that will stimulate germination and then providing the necessary light, water, nutrients, and space for them to grow until you harden them off enough to transplant outdoors.

Let me emphasize that again: plants that have been growing indoors (under lights, the window sill is rarely enough light to grow healthy seedlings) HAVE TO BE HARDENED OFF to survive outside. I will have a short section on hardening off seedlings grown indoors.

Direct seeding means:

Planting seeds outdoors, directly into the soil and location where you want them to grow in your garden.

When deciding which seeds to start indoors you need to consider:

1. How large will the plants get before transplanting?
2. How much space do you have for growing plants indoors?
3. What are the benefits of having a more controlled germination environment?
4. What might be so small upon sprouting outside that it might be easily lost or covered up?
5. What plants might suffer more from transplant shock?
6. How long does it take a plant to reach desired maturity?
7. Will the plant be hardier if direct seeded?
8. Are there pests or diseases that the new seedlings will be vulnerable to in either situation?
9. How impatient are you to start new plants?
10. Are you growing a number of plants suitable for your garden size?
11. How long can the plant be optimally grown in a pot before transplanting?
12. What is the expected germination rate?
13. How expensive is the seed?
14. Is it a cool weather or warm weather plant?

15. How reliable is your water supply in the spring in case there is not enough rain?

16. What will keep you from impulse buying from the garden center?

Remember that there is always some shock or adjustment time for a plant when it is transplanted. How long it takes the plant to recoup from that will depend on things like how much the roots are disturbed and how dramatic the weather is. I always assume it will take about two weeks for the plant to be in full growing mode again.

Understanding average last frost dates

A frost can happen if temperatures are close to freezing. The average last frost date is the dividing line between crops that can grow in cooler temperatures and those crops that will be severely harmed or killed by a frost. There will be nuances on this as various plants are discussed.

An old time maxim for gardeners in southwest Idaho has been that danger of frost is past when all the snow is melted off of Bogus Basin, a local mountain. That is usually by the first weekend of June.

Since the powers that be starting keeping track of weather patterns, that has been adjusted. That is, of course, affected by how long of a time period is used to gather dates of last frost. The date that I have seen given many times is May 15.

But, remember, that is the average.

Some springs are warmer sooner, causing plants in pots to struggle in the heat. In those cases, I err on the side of planting them out sooner, but keeping a close eye on nighttime forecasted lows. I cover my plants anytime the forecasted low is 40°F or lower because too many times I have seen the forecast be off by 10°F.

If the temperatures seem to be staying cooler longer, my main concern is if the nighttime lows are too far below freezing. While you can protect from a light frost by providing a temporary cover, it will not be enough protection against a deep freeze. In my experience, the milk carton mini-greenhouses will work down to 30°F in most cases. It will depend on other factors, too.

For instance:

1. How warm was the day before the frost? Can you time covering your plant to trap in a bit of heat?

2. What is your micro-environment? Not all locations get as cold as the main forecast.

3. What moisture is involved? Moving moisture can guard against frost, as orchard growers know. However, static moisture can cool things off.

Part Four

Seeds I Always or Almost Always Start Indoors

Tomatoes

Though I have had tomato plants volunteer in my garden (grow from last year's debris without any help from me), tomatoes are easy to start indoors and fun to put out into the garden as plants. I like to plant them indoors around late February or early March, because I'm always experimenting with how mature I can get the plants before planting them out, but I've planted them in the greenhouse as late as the first week of April and had healthy plants for transplanting after the last frost.

If I was going to direct seed tomatoes, I would put the seeds in the ground sometime in the last couple weeks of May or in the first week of June. That is when the weather will be reliably warm enough.

Peppers

I have never had any peppers sprout in the garden from last year's crop. Also, peppers, especially the hotter varieties, seem to need more time to produce fruit than our average warm weather season in southwest Idaho, so giving the plants a start in the greenhouse should give me more of a chance of reaching that phase of growing. I really like to get my peppers sprouting by late February. They seem to take a bit longer than most vegetable seeds to sprout.

Strawberries

I have only grown strawberry plants from seed a couple of times, but the seeds are extremely small and expensive, so I choose to keep track of them in pots in the greenhouse. For the record, I have also planted several times using crowns, which is what the roots of strawberry plants with a bit of dormant and trimmed top growth are called. I have also propagated my strawberries from the runners. Those last two ways are probably much more reliable than planting from seed.

Oregano and Thyme

A lot of herb seeds, like oregano and thyme, are very small and the plants take a long time to reach a size where I can keep track of them outdoors in the general garden. Many also take a long time to germinate, which is easier for me to keep track of indoors. Dill and basil are covered later because they are easy to direct seed.

Petunias and Violas

Petunias and violas also produce very small seeds and small sprouts. Plus, I'm always trying to get blooms sooner. I try to get these planted in my greenhouse in February as soon as I can.

Thunbergia

(aka Black-eyed Susan vine)

Thunbergia seeds are somewhat unpredictable in their germination and not very many seeds come in a package, so I start them indoors. Once they sprout, the plants are easy to grow. I usually plant the seeds in pots around early March.

Part Five

Seeds I Often Start Indoors

Lettuce

There are two reasons for starting lettuce indoors. One is that we can actually harvest leaves for eating from the greenhouse. I don't usually start lettuce any earlier than January because of the busyness of the holiday season.

Lettuce is easy to transplant and can also be one of the first things to germinate directly in the garden in late winter (anywhere from late February to early March, depending on the weather). I'm usually planting it from seed outdoors at the same time that I'm transplanting what I've grown in the greenhouse.

Asparagus

I have successfully started asparagus seeds both indoors and outdoors several times. However, the seedlings are small and feathery. In my experience, they have been easy to lose in the garden, sometimes seeming to hide in their own shadows. Because of this they have gotten accidentally grabbed when weeding a nearby plant. Thus, I prefer to start these seeds indoors.

Broccoli and Cabbage

Although broccoli and cabbage germinate very nicely outside, they are easy and fun to start from seed. And like lettuce, I can then plant green, growing plants outside sooner. Since they are frost hardy, they can be transplanted out as early as mid-April.

(Frost hardy plants aren't killed by frost. I recommend the book *How Plants Work: The Science Behind the Amazing Things Plants Do (Science for Gardeners)*, by Linda Chalker-Scott, if you are interested in why some plants can handle frost or freezing and others can't)

It seems to be best if these seedlings can grow for 4-6 weeks before being transplanted.

If I don't get my broccoli or cabbage seeds germinated in time to transplant the seedlings out sometime before the last week of April, I will just wait and direct seed them at the end of April or in the first week of May.

Celery

I haven't grown a lot of celery, but what I have grown I have started indoors around early March.

Basil

Basil is a warm weather plant, so planting it indoors may produce an earlier crop. I start my basil indoors about half the time. Since it is killed by frost and won't thrive in cool weather, I usually wait to plant any seeds indoors until early April.

Nasturtium

Although nasturtiums can be direct seeded, I seem to have better success germinating them indoors. They sprout with relatively large leaves, so that is always rewarding. However, they also grow fairly quickly, so I don't usually start them until about 6 weeks before the last likely frost. That means starting the seeds around the first part of April.

Part Six

Seeds I Rarely or Never Start Indoors

The following seeds either get too large too quickly, have no advantage from starting early indoors, or obviously thrive better when directly planted outdoors. The root crops, in particular, often don't handle transplanting well. Onions are an exception. I have never had any trouble transplanting onion plants.

I will put the plants in groups according to the planting time. Remember I have covered some of the plants that can be direct seeded at these times in previous sections.

Early cool weather crops

These seeds can usually be planted directly from seed as early as the ground can be worked in the late winter or early spring.

Spinach
Radishes
Peas
Cilantro

Later Cool Weather Crops

These can usually be planted from seed sometime in April, being somewhat frost tolerant, but not growing well until there is a bit more warmth and sunshine.

Beets
Carrots
Onions
Dill

Late spring crops that can handle a touch of frost

These can usually be planted late April through early May. You may have to keep an eye on an unusually hard late frost, but they can usually handle a nip of frost here and there.

Sunflowers
Potatoes *(typically grown from pieces of potato)*

Crops that need warm weather and will die in a frost

These seeds want the ground warm. And even though they need water to germinate, they do not like soggy soil. I don't usually put these seeds in the ground until the last couple weeks of May at the earliest. You can wait until the first week of June and get good crops from most of them.

Beans, bush or pole or dry shelling
Corn
Cucumbers
Melons, such as cantaloupe and watermelon
Winter squash, like pumpkins and acorn squash
Summer squash, like zucchini and yellow crookneck

Part Seven

More on Flowers

The following are annual or biennial flowers that I grow regularly from seed. If there is a flower you like a lot, I recommend watching for volunteers the next season and you will get a better idea of when it likes to germinate in your environment. For instance, even though the packages of sunflower seeds say don't plant until after danger of frost is past, I always have sunflowers sprouting in late April from seed that fell from the previous year's flowers. Also, some seeds germinate in waves, so if the weather turns unfavorable, others of those seeds germinate later and there is another chance for them to propagate again.

Cool season flowers

These plants either thrive in or tolerate cooler temperatures

Hollyhocks -	usually direct seed
Sweet Peas -	direct seed
Tidy Tips -	direct seed

Columbine -	direct seed or start indoors, will volunteer and spread if in a favorable environment
Bachelor Buttons -	direct seed
Double Shirley Poppies -	do best with direct seeding
Snapdragons -	easy to direct seed
Painted Tongue -	have only ever tried starting indoors
Mission Bells Poppy -	direct seed
Marigold -	usually direct seed

Warm weather flowers

These plants are either killed or seriously harmed by frost or freezing.

African Daisies -	start indoors
Zinnias -	usually direct seed
Heavenly Blue Morning Glory -	usually start indoors, but frequently have volunteers
Moss Rose (aka Portulaca) -	start initially from seed, but then get lots of volunteers
Cosmos -	best luck with direct seed

Zonal Geraniums -	have only started indoors, was surprised by a couple of volunteers last year
Dahlias -	grow most from purchased or saved tubers, but have grown some varieties from seed
Rudbekia -	start indoors, but get quite a few volunteers in some places
Four O'Clocks -	direct seed, very vigorous self-sower after that

Part Eight

Seed Starting Basics

General seed starting basics

Seed depth - The pre-plant that is tucked inside a seed starts at a predetermined size. Upon sprouting, it needs to be able to reach the surface of the soil. This allows it to access light and air in order to keep growing. If it is planted too deep, the poor little sprout will be buried the same day it sprouts. I use the rule of thumb of planting a seed to a depth twice the diameter of the seed.

If a seed is planted too shallow, it will not have good support when it sprouts. The risk will be that the top of the seedling will over balance the roots, which might not have enough soil to grab. If the roots end up on the surface of the soil, the seedling may also be inclined to bend in the wrong place.

Dark or Light - Here we are just talking about germination. All seedlings need light as soon as they are germinated, but some seeds need dark in order to germinate and some need light.

Smaller seeds, particularly flower seeds, may need light in order to germinate. Some seeds won't germinate unless kept in the dark, even if only buried under ⅛ inch of soil. These things are usually mentioned clearly on the seed packet if they are important. If they are not mentioned, it probably doesn't matter or planting depth deals with it. Obviously, the deeper a seed is planted, the more dark it will be.

Temperature - Soil temperature is important, even if it is dark. Most seeds have a temperature range in which they will germinate best. These temperatures seem to be related to the type of weather that the mature plants thrive in.

Evaluation - If you really want to monitor what is going on with your seeds, consider planting enough seeds to incorporate some research into the process. Possibly have the same type of seeds set to sprout in a paper towel in baggy so that you can easily monitor progress. Make sure to keep them in the same temperature as those in the pots. This will not exactly duplicate the environment in the soil, but it will come close.

One more option would be to use clear plastic cups for some of your pots (Don't forget about arranging drainage). Plant the seeds right next to the edge, where you can observe them more easily. This is not usually the recommended location for seedlings you hope to transplant because it means a lopsided root mass, as well as putting the stem more at risk during transplanting. However, learning what is going on with various seeds can make you have more success with your seed germination in the future.

Seed starting basics for In the greenhouse

Besides seeds, you will need:

Pots or containers - Although any container will do for a pot, it can be challenging to create adequate drainage in homemade pots. The main issue is making drainage holes large enough holes that don't destroy the bottom of the container. It can also be hard to make pots that have a ratio of height to depth that both allows for efficient use of potting soil and easy water management. If all the water is draining to the bottom of an over long pot, it doesn't do much good for a new seedling near the top. If there isn't enough depth of soil, water retention will be a problem.

Biodegradable pots are not always as biodegradable as advertised. I have seen transplanted seedlings that never thrive because their biodegradable container was still intact enough to constrain the roots.. Also, the top rim of a biodegradable pot can restrict the plant, even if the bottom of it mostly dissolves. Some of these problems may be because of the relatively dry climate in southwest Idaho. When I tried using biodegradable pots, I ended up just removing my seedlings from them for transplanting.

Take care at the time of transplanting to make sure the top edge of the potting soil in the pot is even with or just below soil level you are planting in outside. Avoid the other extreme, too, so the plant does not get planted too low into the ground for its liking. Something like tomatoes can handle being planted higher up the stem, but not all plants can.

If you buy any potted plants, whether pony packs or larger, you can always save the pots. I have done this for years, recycling them until they disintegrate. I also have a variety of pots specifically produced for the home gardener. Most of them have lasted for years with little sign of wear, so I consider them worth the expenditure.

Choose your pot size based on how big the seedling will grow in the first 4 weeks. This will allow for space efficiency, while giving the seedling room to grow. It might make it to 6 weeks, but many will benefit from repotting past that time period. Remember, the pot confines the roots, which in turn limits access to water and ability to support top growth. If the roots get to entwined, or “root bound,” in the smaller pot, the plant may never recover.

Fill the pots with soil up to about ¼ inch to ½ inch below the top edge of the pot. If they are filled to the top, they will be harder to water, as the water will tend to flow off before it has much chance to soak in. Also, there is a higher risk of the seeds being washed away.

You will probably also want some sort of trays to help the pots sit on the shelves and to making moving them around easier. The best trays I have found are from Territorial Seed Company.

Soil suitable for germination - A good quality potting soil is recommended. It doesn't have to be specifically for seed starting, just not too chunky. The quality of the potting soil matters because the lower quality types tend to get hard when they dry, much like regular dirt from outside does. It can be very hard to get such soils to reabsorb water after they have dried past a certain point. Exact color doesn't seem to matter, although most of them will be a pretty dark brown. The common wisdom is that darker soil is healthier, but I have found color of potting soil to be an unreliable indicator of quality. The best potting soil will feel loose and somewhat fluffy, but not sandy or dusty.

It is possible to make your own potting soil using a 1:1:1 mix of compost, peat moss, and something like perlite or vermiculite. Another mix recommended by *The New Seed Starters Handbook* is 1:1:1 compost to loose garden soil to vermiculite.

To prepare whichever soil you use for seed starting, mix in enough water to make it the consistency of thick oatmeal. One of my guidelines is to make it wet enough that just a bit of water can be squeezed out of it. If there is a little water pooling in the bottom of the bowl when you get down there, just add a bit more soil to get the correct mix again.

After putting the moistened soil into a pot, tamp it down some with your fingers to get rid of pockets of air. As you do this, even out the top of the soil. It will be easier to plant seeds at the appropriate depth if you do these things.

Grow lights - When people say grow lights, it also implies the whole set up of lights and shelves, but we need to talk about the components separately.

I purchased bulbs at a local DIY plumbing and electrical supply store. I got warm color, 3000K bulbs and cool color, 5000K bulbs, both with color rendering of 85%. It seems backwards that the cool colors have the higher temperature, but my engineer explained it to me. The lights are only called cool or warm according to the hue of their light because that is how most people think of those colors. Blue is hotter. You need the different types of light bulbs to achieve the spectrum of light that most plants need in order to grow.

We got the shop light fixtures at Home Depot. We attached them to typical rolling shelves you can get almost anywhere. A more detailed description of how to do this will be the subject of a later booklet.

Air flow - With all the lights on and the humidity from watering, a greenhouse can get stuffy. I keep my eyes out for sunny afternoons when I can crack a window or two without lowering the temperature so much that it might shock the seedlings. So far I have not had to resort to fans. It also helps that I start hardening off my seedlings as soon as the weather permits.

A way to water - I have a coiled hose and a spray nozzle with both a misting and a shower option among the dial settings.

Labels - I usually use masking tape on the side of the trays when the seeds are still sprouting. When the plants are taller and I am likely to move individual pots around, I use jumbo craft sticks written on with Sharpie marker.

A place to harden off seedlings - When the time comes you will need a safe place to set your trays and pots in order for them to get gradual exposure to the outdoor elements. Ideally this will be a place close by and where you can easily get them inside if there is a sudden change in weather that might harm them.

Fertilizer - If the seedlings are in pots for more than about 4 weeks, you will want to fertilize most of them for healthy growth. This will depend on whether or not you purchase potting soil that has a slow release fertilizer. Seeds don't need fertilizer to sprout.

Plastic bags (optional) - I use large clear trash bags as mini greenhouses for the first 2 - 3 days to help the seeds stay more moist.

Seed starting basics for outdoors in the garden

Besides seeds, you will need:

Soil - This can usually just be typical top soil amended with a 1-2 inch layer of compost every year. It doesn't need to be fancy or perfect right away, though it will get better over time as you keep working with it. However, you don't want it to be either extreme of hard packed clay or loose sand. It needs to be loose enough for seeds to sprout *and* have enough substance to hold some water without the water all draining right away. Usually a little tilling the first year does the trick, unless your home is newly constructed and all the topsoil has been bulldozed away.

Some seeds are more temperamental than others. Carrots, for instance, are very particular about the soil on top of them being light and loose, so I usually only cover them with a thin (⅛ inch) layer of my seed starting soil.

Watering system - Most seeds need very regular watering to germinate. Think spring rain. I don't find drip lines to work well for seeds that are cool weather crops. Warm weather crops seem to do fine with drip line watering. Your life will be much more enjoyable if you have your watering system figured out before you plant your seeds.

Sometimes it rains enough in the spring to do the trick, but southwest Idaho is high desert, so it is likely you will need to do some watering yourself. Since most irrigation/sprinkler system water is not turned on until there is no danger the pipes can still freeze, hopefully you have a faucet from the house that you can use for sprinkling your early seeds. If not, there are some nice watering cans, although that is obviously more labor intensive.

Frost protection - If you want to plant frost sensitive plants out when there is still a danger of a light night time frost, then you should have some way to easily cover the seedlings on evenings when the forecast low is 40°F or less. Although you can use any container that will fit over the top of the seedling without breaking it, my favorite is plastic milk cartons with the bottom cut off (and the lid still on). I do NOT recommend metal containers, as they transmit the cold inside much faster.

<u>If you cover plants like this, you must be diligent to remember to take the plastic containers off before the next day's sunshine cooks the plants</u>. This can happen very quickly once the light of the day hits. Set an alarm to remind you!

Part Nine

Hardening Off Your Seedlings

Why do you need to harden off your seedlings?

It is very important that you harden off any greenhouse seedlings before planting them permanently outside. This is also true for any purchased nursery plants that have been sheltered from direct sunlight. It's kind of like a tan versus a sunburn, only it is much more life threatening for plants.

Even though plants can grow well under greenhouse lights, the strength of those lights is not anywhere near the intensity of sunlight. Thus, in the greenhouse, their leaves grow to take in as much of this light as possible. This means that leaves that grow in the greenhouse, or under shelter, will be shocked or burnt when all of a sudden subjected to long periods of direct sunlight. Without functional leaves, a plant will die.

If you grow the exact same plant in the greenhouse and outdoors, you can see the difference in the leaves. Usually the outdoor plants will seem smaller, stockier, and darker. The greenhouse plant leaves will be broader, lighter green, and even feel more tender.

Sunlight is not the only environmental factor that the plants need to get used to with hardening off. They also need to toughen up to wind. It's kind of like developing their abdominal muscles from trying to keep their balance. In the greenhouse, there may be fans for air movement, but there is nothing like a stiff breeze or the nearly continual winds that can blow all day outdoors.

How to go about hardening off your seedlings

To prepare a greenhouse seedling or plant for life in the wild outdoors, it must be gradually hardened off. This can take anywhere from a couple of weeks to a couple of months, depending on the weather and how consistent you can be in the process.

For the first days of hardening off, I try to watch for very mild, sunny weather. I don't want the plants to be subject to too much of a temperature difference, but I don't want it scorching hot, either. It's is generally about 75°F my greenhouse, so about 60-65°F is what I aim for outside.

A little bit of overcast can be useful, as long as it isn't too cold and blustery. The best first hardening off time is most often going to be an afternoon event for hardening off anything in March or early April here in southwest Idaho, which is when I begin for some things. In late April and or early May, I tend to utilize morning hours more.

I use a timer faithfully if my plants are only ready to spend a few hours outside. Their lives depend on it. Then, I don't turn it off until I am actively attending to relocating my seedlings. If I need a few more minutes to get to it, I set the timer again. It is too easy to lose track of time while doing something else.

I usually harden my seedlings off in increments of hours. That is, I will put them out for one hour the first day, then maybe two hours the second day. But I don't necessarily always increase the amount of time. I will often put them outside for the same length of time for 2-3 days in a row. I consider them tender and vulnerable until they are up to 6 consecutive hours in one day, but usually still bring them in at night until they have been outside for 6 hours for at least three days in a row.

It is not necessary that the plants go outside every day for the hardening off to be effective. If there is a day or two where the weather or your schedule are not cooperative, simply start at the same amount of time or slightly less than the last time you took them out. It all adds up to strengthen the plants.

When you don't have grow lights, but still want to plant seeds indoors

Another variation to hardening off seedlings is to put them outside once they begin to sprout. Sprouting in pots is usually more productive if they are kept at constant temperatures and moisture. This is easiest to maintain indoors in the earlier parts of spring for some plants. However, they can handle the sunlight from the get-go <u>if you get them out there **as soon as** they show their pretty little green faces.</u> Keep in mind, though, that they are still very vulnerable to drying out. Think of watering them like gentle spring rains, especially until they have more root and leaf growth. Also, the outdoor temperatures need to be consistently within a range that the plants will grow in.

Other things to think about when hardening off your seedlings

Sometimes I will put more tender plants in a place that will get some dappled sun or shade sooner than the others, particularly if it is a bright, sunny, warm day. If you study the patterns of sunlight during the day in your preferred hardening off zone, you will likely discover micro-environments that can work to your plants' advantage.

As the time spent outside gets longer, closer attention needs to be paid to water needs. Not only does the direct sunlight lead to more evaporation, but the plants usually have more foliage, which uses up the water faster. Since they are still in relatively small pots, they need to be monitored. Looking at the soil is not a good way to judge how moist the soil is. You really need to feel the soil down to ½ to one inch depth.

One thing that has to be guarded against is unexpected downpours. The force of such water can wreak havoc on small seedlings in general, but it also can displace soil from the pots or threaten to drown them because the pots cannot drain quickly enough. If there is any chance of such a rain while I am gone from the house or otherwise unable to keep a close eye on my plants, I put as many plants as possible under the eaves of the house for protection, but where they still get sunlight. I have learned the hard way, though, to make sure they are pushed far enough under so that gutter overflow doesn't cascade down onto my plants.

If my plants do get subjected to a hard rain, it is important to dump any extra water out that has collected in their trays. Even if the pots can theoretically drain, they won't if the tray is full of water. A tray can collect a lot of water in a good rain, which can take too long to evaporate for the health of the plant.

Once you have reached the 6 hour mark

Once the plants are up to 6 consecutive hours outside during the day for at least three days in a row, they can handle some warm nights outside, too. I use 40°F as my benchmark nighttime low for this, but also evaluate wind chill. Keep in mind that because my patio is south facing, it probably actually stays warmer than that. Plus, being close to the house and somewhat protected from the sides, there is a pocket of warmth there.

When the nights are warm enough, I can leave the plants out 24/7, but I watch the nighttime temperatures diligently. I haven't spent all that effort growing those plants just to lose them in one cold night! I also watch for other weather factors and place the plants in more protected places as seems wise, i.e. under the eaves or benches, to be able to sleep more peacefully.

You can see there is no secret science to the process. It is just a matter of diligence and monitoring. Taking time to properly harden off your plants will not take very much effort compared to the whole of the gardening season and will be well worth doing right. Soon you will have plants that are ready to live the rest of the season outdoors.

Part Ten

Transplanting Your Seedlings

There are a few basics that will go a long way toward successful transplanting of seedlings.

Choose the best or correct season for the plant to grow.

For instance, you can plant tomato plants out in early April, but they probably won't thrive and you will need to do a lot of work to protect them from freezing temperatures. Many times I have witnessed tomato plants that are transplanted out right as the temperatures are reliably warming up easily catch up with tomato plants that have been outdoors in the cold for a few weeks longer. On the flip side, lettuce will do best when transplanted during the cool season. If you wait until warmer weather, the lettuce will bolt *(go to seed)* more easily. Also the heat tends to produce bitter leaves if lettuce doesn't get enough water.

Try to do your transplanting when the plant will be least stressed by the weather.

That means looking for a cloudy day or transplanting in the evening. This gives the plant a few hours to establish its roots before it is tested against the elements.

Treat the stem with care

Your beautiful little seedling can more than likely handle a leaf getting torn or getting a few roots get nipped. What it can't survive is the stem getting broken. When I am transplanting, I support the seedling in one of three ways.

One way is to hold the pot upside down on the palm of my hand, with my fingers carefully on the top of the soil on either side of the stem until I'm ready to set it into its hole.

Another way is to just hold the soil-root mass in my hand or a shovel until I plop it into its hole.

A third way is to hold the seedling delicately by a leaf. This is only a good way if there is no weight of soil on the roots.

<u>The main thing is to not hold it by the stem</u>.

Dig the hole deep enough for the roots

This is more straightforward if you have enough roots so that they are holding some dirt around them. If the dirt falls away from the roots, there is no need to panic. Simply hold the seedling gently by the leaves as you place the soil around it in the hole. While the root will probably grow down sooner or later, it will be more at risk if it is unnaturally left closer to the surface, so I take care to let it hang down like it was before the soil fell away.

Stimulate roots to grow outside the shape of the pot

If the roots have grown to hit the outside of the pot and then turned in again, they will benefit from some rubbing or slight trimming to signal to them to grow in a different direction. Without this the roots are at risk for not growing out into their new soil, which will stunt overall growth.

Water them to help establish them

This means two things. First, watering the seedlings right away helps the soil to settle around the roots and make contact. Second, until they have developed more roots, the seedlings will need watering similar to how they were watered in the pots. That means watering relatively more often until they have the root system to support them in getting water from a greater area around them. This will be weather dependent.

Trim off blossoms

If your plant has buds or blossoms, prune them off when they are still in the pots and right before transplanting. We are not talking about tulips here, haha. We are talking about vegetable seedlings that will produce multiple blossoms and fruits over the season. Cutting off the blossoms will relieve the plant of putting energy toward producing fruit while it is establishing itself in its new environment.

This only needs to be done before transplanting. Once the plant is in the ground, it will probably not flower again until is more established.

Monitor your new transplants

Seedlings are vulnerable. Check on them more often than larger, established plants. Whether it be insects, stress from the weather, or your neighbor's cats, you will have a better chance of saving them if you catch a problem sooner.

Seed Planting Guide Chart

Keep in mind that this chart is a guideline. Weather conditions vary from year to year. Microclimates around your garden may impact your results. Also, time it takes a plant to reach maturity can be different for varieties among the same plant type. This can affect when you might want to start seeds indoors.

Since I've only charted by months, and not days within the months, you will need to refer to specifics in the book for whether this is earlier or later in the month. Then, maybe you can color code your chart to indicate this.

Type of Seed	*indoors*	*outdoors*	*Ja*	*F*	*M*	*Ap*	*M*	*Jn*	*Jy*	*Au*	*S*	*O*	*N*	*D*
asparagus	*O*	*O*	*IP*			*OP*								
basil	*O*	*O*				*IP*		*OP*						
beans	*RN*	*A*					*OP*							
beets	*RN*	*A*				*OP*	*PA*	→	→	→				
broccoli	*O*	*O*			*IP*	*OP*				*IP*	*OP*			
cabbage	*O*	*O*			*IP*	*OP*				*IP*	*OP*			
carrots	*RN*	*A*				*OP*	*PA*	→	→	→				
celery	*A*	*RN*			*IP*		*OP*							
cilantro	*RN*	*A*			*OP*	→	→	→	→					
corn	*RN*	*A*					*OP*							
cucumbers	*RN*	*A*					*OP*							
dill	*RN*	*A*			*OP*	→	→	→						
lettuce	*O*	*A*	*IP*	*OP*	→	→	→	→						
melons	*RN*	*A*					*OP*							
onions	*RN*	*A*				*OP*								
oregano	*A*	*RN*		*IP*			*OP*							
peas	*RN*	*A*		*OP*										
peppers	*A*	*RN*		*IP*			*OP*							
potatoes	*RN*	*A*				*OP*								
radish	*RN*	*A*				*OP*								
spinach	*RN*	*A*		*OP*										
squash, summer	*RN*	*A*					*OP*							
squash, winter	*RN*	*A*					*OP*							
strawberries	*A*	*RN*		*IP*		*OP*								
thyme	*A*	*RN*		*IP*			*OP*							
tomatoes	*A*	*S*			*IP*		*OP*							

Key

A = always or almost always
O = often
RN = rarely or never
S = sometimes
PA = plant again (another crop)
IP = indoor planting time
OP = outdoor planting/transplant time

Appendix

Seed companies I like:

Territorial Seed Company (territorialseed.com)

Baker Creek Heirloom Seeds (rareseeds.com)

Johnny's Selected Seeds (johnnyseeds.com)

Swallowtail Garden Seeds (swallowtailgardenseeds.com)

Select Seeds (selecseeds.com)

Pine Tree Garden Seeds (superseeds.com)

About the Author

Laura Blodgett runs around barefoot in Idaho as much as she can. She and her husband homeschooled their 7 children and are now enjoying being grandparents. Laura doesn't know the meaning of the word bored and wonders how much a person could learn if life spans were as long as they were before Noah's flood!

You can find out more on her author website, laurablodgett.com, where you can also sign up to learn more about her upcoming books. Her other websites are:

thehappyhomeschool.com
dailyimprovisations.com (homesteading blog)
funfitnessafter50.com
funlearningchinese.com

Other books by Laura Blodgett:

Melody's Life Savings - If you want to understand how to have hope that makes a difference in the face of death and where to find such hope, read or listen to *Melody's Life Savings*.

How to Build a Backyard Brick Oven From Scratch - Have you dreamed of having your own backyard brick oven? See how one man built his from the ground up using basic materials!

Why Does Grandma Run Barefoot? - What if there was one simple way you could feel younger and more energetic? It can all begin with learning how to use your feet right again! After over 10 years of running barefoot, Laura is more active at the age of 58 than she was in her 20's. She wants to share how this is possible for you.

Children's Family Read-Aloud Picture Books:

The Blue Fish (also in Chinese) - Enter into the wonder of a child enjoying her grandfather's blue fish in the pond.

A Million Rocks (also in Chinese) - Explore many almost counting words on a trip to the beach with the twins. Almost counting words are used all the time in every day conversations, but do you know what they mean?

Four Frogs in a Glass Pond - The children were thrilled to be on a fishing trip with their dad. But they never expected the worms they cast into the mountain pond to bring back frogs. And when the stubborn hoppers wouldn't let go of the line, they opened up the door to a whole new adventure.

Junior Novels:

Lulu and the Big Kaboodle (An Adventures with Lulu Book) - *(for 7-12 year olds.)* Lulu is having misgivings. Because strange dogs had been mean to her before, she didn't really like them. But she still decided to say yes when her little girl asked her to babysit a big, black, crazy puppy.